GUESTS OF TIME

Guests of Time

Poetry from the
Oxford University Museum
of Natural History

edited by JOHN HOLMES
with photographs by SCOTT BILLINGS

VP

Valley Press

First published in 2016 by Valley Press
Woodend, The Crescent, Scarborough, UK, YO11 2PW
www.valleypressuk.com

ISBN 978-1-908853-80-6
Cat. no. VP0097

A CIP record for this book is available from the British Library.

Cover and interior design by Jamie McGarry.

All photographs by Scott Billings except:
page 32, 'Sea Spider' (Image credit: Professor Derek Siveter)
page 66, 'Red Lady' (Image credit: Sarah Joomun)

Contents

Preface — PAUL SMITH 7
Introduction — JOHN HOLMES 11

In the Grand Concourse — JOHN BARNIE 19
The O'Sheas — STEVEN MATTHEWS 20
"Boughs being pruned" — GERARD MANLEY HOPKINS 23
A Vision of Linnaeus — JOHN LUCAS TUPPER 25
The Garden — WALTER DEVERELL 27
Argonauta — KELLEY SWAIN 29
Ammonite — STEVEN MATTHEWS 30
Rorqual — KELLEY SWAIN 31
The Sea Spider Speaks — JOHN BARNIE 33
To the Palaeontologists — KELLEY SWAIN 34
The Crystallographer — STEVEN MATTHEWS 35
Corsi Marbles — STEVEN MATTHEWS 39
Double Refraction — KELLEY SWAIN 43
A Year and a Day — ELIZABETH SIDDAL 44
Summer Dawn — WILLIAM MORRIS 47
The Woodspurge — DANTE GABRIEL ROSSETTI 48
A Quiet Place — WILLIAM MICHAEL ROSSETTI 49
Daphne, The Orchid Mantis — KELLEY SWAIN 51
Flashman — KELLEY SWAIN 52
Who Shot Cock Robin? — JOHN BARNIE 54
Let's Do It — JOHN BARNIE 55
The Wandering Albatross and the Storm Petrel — STEVEN
    MATTHEWS 57
Shadows — JOHN BARNIE 59
Emblems — THOMAS WOOLNER 60

"Yet With Time's Cycles Forests Swell" — STEVEN MATTHEWS   62

Here's A Modest Apology for Anthropo-Pithcology — JOHN
    ADDINGTON SYMONDS, SR.   64

An Old Gordian Knot — JOHN ADDINGTON SYMONDS, JR.   65

A Welshman Looks at the Red Lady — JOHN BARNIE   67

Fibonacci Crystals — STEVEN MATTHEWS   68

The Spirit Collection — JOHN BARNIE   71

The Spirit Room — STEVEN MATTHEWS   72

Lyme — KELLEY SWAIN   74

Cockroaches On My Mind — JOHN BARNIE   75

The Lay of the Trilobite — MAY KENDALL   77

Oxford — KELLEY SWAIN   81

Notes   83

Biographies   87

Preface

Oxford has an important place in any historical account of the development of modern science. The philosopher-monk Roger Bacon, the *Doctor Mirabilis* of thirteenth-century thought, helped to lay the groundwork for a philosophy of science that underpins the modern method. Four centuries later, during the civil war and the later Protectorate, the Oxford Philosophical Club began to meet in Wadham College under the leadership of the Warden, John Wilkins, attracting others, such as Boyle, Hooke and Wren, into the fold. With the reinstatement of the monarchy, it was this group that returned to London to found the Royal Society in 1660.

However, by the mid-1800s Oxford was far from being at the cutting edge of scientific endeavour, and it was clear that it was falling behind other centres such as Edinburgh, London and Birmingham. Laboratory resources and infrastructure for experimentation were scant. The old Ashmolean Museum (now the Museum of the History of Science) had been designed in the 1680s to have laboratory facilities in its basement and Christ Church had its own laboratory from 1767 but the provision was haphazard. Charles Daubeny, a chemist and botanist, even went to the lengths of building a laboratory at his own expense at Magdalen College. Out of this scatter-gun approach emerged a growing campaign for a modern facility that would bring together scientific and medical research across the university, with Henry Acland in the vanguard.

In more pragmatic times a laboratory facility for the University of Oxford might not have resulted in the University Museum, but Acland's vision was comprehensive. A museum would lie at its core, in the centre court, and the science departments would wrap around the cloisters. Acland had been a close friend of John Ruskin since their days as students at Christ Church so it was only natural that Ruskin should be approached for input to the design, as John Holmes explains below. The original concept for the University Museum, now the Museum of Natural History, provides unification at a number of levels. For the first time since the Oxford Philosophical Club, the sciences were brought together, with collections, research and teaching under a single, spectacular roof. At the same time, Acland's view was that this should also be a place that lay at the meeting-point of art and science. The result is a building that is both unique and overwhelming in its detail.

It would be easy for a museum such as this to become overly self-referential about its history, and to become a museum of itself. Yet it was designed as a place of contemporary science and we remain committed to research and to engaging and involving the widest possible audiences with the sciences of the natural environment. Temporary exhibitions in our Contemporary Science & Society series explore a variety of themes at the cutting edge of modern science, particularly where they have an element of societal relevance, dispute or debate. In addition, the museum has used 2016 to examine once again the interface between art and science under the banner *Visions of Nature*. This has encompassed the visual art of Kurt Jackson for his exhibition *Bees (and the odd wasp) in my bonnet*, together with *Microsculpture*, the giant and striking images of insects from the museum collections by photographer Levon Biss. The third element of the *Visions of Nature* year is this volume. Through 2016, we have hosted not one but three poets-in-residence with the hope that they would intersect, collaborate and explore every corner of the museum and its collections. John Barnie, Steven Matthews and Kelley Swain have become a familiar sight around the museum and delving into collections drawers. We are immensely grateful to them for their work, but also for making us look afresh at our familiar collections.

The results are here, interleaved with work by nineteenth-century poets linked with the museum selected by John Holmes and photographs of the museum and its collections taken by Scott Billings. We hope that you enjoy this collection as much as we have enjoyed hosting the three poets, and that you will also take the opportunity to visit the museum and view it with fresh eyes.

PAUL SMITH
Director, Oxford University Museum of Natural History

Introduction

The week before the foundation stone was laid for the new Oxford University Museum of Natural History in 1855, the University's Reader in Anatomy Dr Henry Acland addressed the Oxford Architectural Society. 'Oxford', he told them, 'was about to perform an experiment'. It was about to test out 'how Gothic art could deal with those railway materials, iron and glass'. The museum was to have a 'roof of glass, supported by shafts of iron'. Its court would be ringed with columns 'composed of variously coloured marbles, illustrating different geological strata and ages of the world', while 'the capitals represented the several descriptions of floras'. The museum was in effect the University's first science faculty. But it was not going to teach natural history and the other sciences only through its collections and laboratories. The building itself was to be a conceptual model of the natural world. The combination of Gothic architecture and modern materials would reveal nature to be at once God's creation and a resource for industrial progress. At the same time, the columns would be object lessons in the properties of rocks, the forms of plants and the habits of the animals living among them.

Acland's experiment worked because he was able to muster a sustained collaboration between scientists and artists. The museum's first Keeper, the geologist John Phillips, along with Acland himself and other Oxford scientists such as the chemist Charles Daubeny and the entomologist John Obadiah Westwood, set out the conceptual scheme for the museum and provided specimens and expertise. The architect Benjamin Woodward, the iron-master Francis Skidmore and the sculptors James and John O'Shea and their nephew Edward Whelan gave their ideas physical expression. Acland's friend the art critic

John Ruskin brought the Pre-Raphaelite artists Dante Gabriel Rossetti, Elizabeth Siddal and Alexander Munro on board as consultants. Munro and the other Pre-Raphaelite sculptors Thomas Woolner, John Lucas Tupper and H. H. Armstead carved the first nine of the statues of scientists that make up a pantheon around the court. Ruskin, Woolner and John Hungerford Pollen contributed designs for windows and the main doorway, William Bell Scott and Pauline Trevelyan for some of the capitals of the columns in the court, while one of Ruskin's protégés, the Reverend Richard St John Tyrwhitt, painted two large murals in the geology lecture room.

It was not just personal friendships and taste that led to the Pre-Raphaelites becoming involved in the design and decoration of the Oxford museum. Speaking three years later, when the building was well underway, Acland explained that, if the museum was to represent nature truthfully it had to represent it accurately. For this reason, he said, 'we have sought to hinder all ornament, unless that ornament be free from vicious carelessness; and to stop all professing transcript of Nature, unless it be painstaking, sagacious, and honest. Herein,' he continued, 'we owe a just debt of gratitude to the young school of Artists, called, half in jest, Pre-Raffaelites'.

The Pre-Raphaelites were known for their scrupulous attention to detail in their paintings. More significantly, just as the Oxford scientists turned to Pre-Raphaelite art for a model for how to represent nature truthfully, so the Pre-Raphaelite artists had founded their own practice on the model of empirical science. In their manifesto magazine *The Germ*, which ran for just four issues in 1850, one of the members of the Pre-Raphaelite Brotherhood, Frederick George Stephens, wrote:

> The sciences have become almost exact within the present century. Geology and chemistry are almost re-instituted. The first has been nearly created; the

second expanded so widely that it now searches and measures the creation. And how has this been done but by bringing greater knowledge to bear upon a wider range of experiment; by being precise in the search after truth? If this adherence to fact, to experiment and not theory,—to begin at the beginning and not fly to the end,—has added so much to the knowledge of man in science; why may it not greatly assist the moral purposes of the Arts?

The following year another P.R.B., William Michael Rossetti, insisted that the Pre-Raphaelites were committed to 'investigation for themselves on all points which have hitherto been settled by example or unproved precept, and unflinching avowal of the result of such investigation'. The Pre-Raphaelites held themselves to the rigorous standards of science, whether they were drawing from nature and or studying human character.

With a symmetry rare in collaborations between science and the arts, Acland repaid the Pre-Raphaelites' compliment to science by following their artistic example. Stephens completed the circuit by reviewing the museum favourably in *Macmillan's Magazine*. He singled out the O'Sheas for special praise. They had 'tenderly, cunningly, and lovingly studied natural forms, and reproduced them with marvelous fidelity and elaborateness'. It was as if they had sought 'in the lush recesses of the neighbouring river-bank for crisp water plants', or in 'shady woods' for 'dank ferns', just as the Pre-Raphaelites themselves would have done. 'Nothing', Stephens concluded, could be 'more intelligently faithful' than their carvings. Combining the O'Sheas' carvings with Woodward's representational architecture, Ruskin's and Pollen's designs, and the statues by Munro, Woolner, Tupper and Armstead, the Oxford University Museum has a good claim to be the single greatest work of Pre-Raphaelite art, as well as being one of the richest and most stimulating collaborations between artists and scientists in British history.

The museum's connection to the arts does not stop with its own building, however, nor with the visual, tactile and spatial media of architecture and sculpture. The Pre-Raphaelites are primarily known today for their paintings, but from the outset they found their voice in poetry. Many of the poems printed in *The Germ*, by the Rossettis, Woolner, Tupper and others, are verbal cartoons for truthful depictions of nature. Others are meditations, sometimes comic, on the implications of natural science. Pre-Raphaelite poetry was crucial to the formation of the aesthetic that was to be set in stone, iron and glass at Oxford. In one case – Tupper's poem 'A Vision of Linnaeus' – a work of sculpture in the museum was completed as a double work of art through a poem written by the sculptor at the same time as he was carving the statue. In its turn, the science taught at the museum and the controversies debated there prompted men and women studying at the University to work through their implications in poems of their own.

The museum opened in 1860, the year after the publication of Charles Darwin's *On the Origin of Species*. One of the earliest and most famous clashes between Darwin's supporters, led by Thomas Henry Huxley and Joseph Hooker, and the old guard of natural theology, led by Samuel Wilberforce, Bishop of Oxford, took place at a session of the British Association for the Advancement of Science, held in the museum's library that year. Darwin's book would initiate a sea-change in science and in humanity's conception of itself. Evolution went from being a minority viewpoint, barely respectable within science, to a mainstream scientific theory. With evolution, we became a part of nature, not apart from it. If natural selection was the engine of evolutionary change too, then we were the product of a brutal, brainless process, not the culmination of God's design. Many of the elder scientists working at the museum, such as

Phillips and Daubeny, resisted Darwin's challenge to natural theology. Others, somewhat younger, including Acland, sought to reach a détente, arguing that evolution was itself the working out of God's plan and downplaying the role of Darwin's preferred mechanism. Still others read the writing on the wall and concluded that the age of human exceptionalism was over, and that we would have to learn to live with our own diminished importance, while recognising how far we were implicated in and subject to the processes of nature.

The science taught by the museum today and the research that continues within its walls are the products of descent with modification from Darwin's own researches. The O'Sheas' carvings too reveal an awareness of ecology that matches his own. Yet the building as a whole, with its Gothic architecture and an angel above the main door, still speaks to us of nature as a divine creation. Through poetry, just as in architecture, we can hold these competing visions of nature in suspense, thinking through them in turn, pondering their relevance then and teasing out what and how they mean to us here and now.

The poems in this anthology are both a tribute to the artistic origins of the Oxford University Museum and a rejuvenation of its artistic legacy. Our title comes from a poem by John Addington Symonds, who studied in Oxford in the late 1850s when the museum was being built. In his sonnet sequence 'An Old Gordian Knot', Symonds describes humanity as 'this momentary guest of time'. Natural history museums remind us just how short a time human beings have been around, and how many other guests of time have been and gone before us. At the same time, they allow us to reach across the aeons, to stage imaginative encounters with relics of the distant past or with the more recent dead. Walking around the Oxford museum, you inhabit several different timespans at once: the hundred and sixty years since the museum was founded; the four hundred years that have gone in to building up its collections; the two-and-a-half-thousand-year history of Western science, embodied in the statues of the scientists round the court; the lifespans and afterlives of the taxidermied animals; the brief but current lives of the bees, cockroaches and other insects who live in hives and cases on the upper floor; the prehistory of our kind, from Australopithecus to the Red Lady; the receding chronology of geological epochs and their diverse life forms; even the cosmic timeframe of the museum's oldest specimen, a meteor that dates back four and a half billion years.

Like the museum itself, the poems in this book span these different times. The majority were written by three contemporary poets who have themselves been guests for a time of the museum. In their previous poetry, John Barnie, Steven Matthews and Kelley Swain have often reflected on humanity's place in and bearing on nature. Over the last year they have got to know the museum with a special intimacy, taking inspiration from its collection, architecture and histories, and discovering their own insights into them and through them into nature at large and human history within in. As well as eight poems by each of them, this anthology includes eleven poems by Victorian poets connected to the museum. Dante Gabriel Rossetti, Tupper, Woolner, Siddal and William Morris were all part of the circle who worked on the museum itself and on the Oxford Union Society, which was built by the same firm of architects at the same time. The poems included in this collection bear on or bear traces of their connection to the museum. William Michael Rossetti and Walter Deverell were further key members of the Pre-Raphaelite group whose aesthetic was invoked by Acland as the model for the stone and ironwork at the museum itself. Gerard Manley Hopkins, John Addington Symonds Jr. and May Kendall were students at Oxford in the early years after the museum opened, each with a keen interest in the science that it was teaching. Symonds's father was a leading Victorian physician who had practiced in Oxford and was connected to the elder generation of Oxford scientists. Together, their poems give us a refracted vision of the museum, its origins and legacy, and of the meanings of natural history, then and now.

Of all the arts, poetry is the most economical and the most precisely incisive. The poems in this collection offer us profound insights into the practice of natural history, both historically and today. More than that, they show us how natural history museums hold up a mirror to ourselves, reminding us of our responsibility towards the things we know and value, and of the need to recall the history of our brief but fraught relationship with the rest of the natural world if we are to safeguard ourselves and our planet even in the short term. Through the scientific study of natural history, we come to know the world around us and the lives of those with whom we share it. Through poetry, we can discover what it means to us to live in this world ourselves, reaching towards the understanding we need if we are to navigate the future with the same degree of confidence as we now know the past.

JOHN HOLMES

In the Grand Concourse

There are crowds under gothic arches,
half a cathedral, half a station,
waiting for the train;
when will it come;
Station Master Darwin might know;
they've all got tags handed out by Linnaeus,
'to each an identity,'
as if hurried from a scuppered Ark;
they say the track is out of line,
they say there's going to be a wreck;
people walk about looking at bones;
when evening comes and lights are turned off
and doors locked,
the station crowd will mill around
in a standing-still sort of way;
millions of years have passed
and they'll never know;
the train is ours now;
I think I see a red light swung
over the future's rails;
there's no screeching of brakes.

JOHN BARNIE

The O'Sheas

As the cart grinds cobbles on Long Wall Street,
water lurches in the terracottas,
brick-red basins knock on one another.
White Alisma flowers sway violently;
curved leaf-ribs, crushed in, lose their proportions.
Now the wind chills them, now the horse stumbles.

James and John dart to the wooden tailgate
together, reach hands over to steady
the load, separate leaves from each other,
return plantain leaf-arches to their grace.
The brothers both nod to the stopped cartman;
the whole rig carefully makes its way on.

Early each day another exotic
parades from Botanic Garden or Parks
through streets to the shell of the Museum.
Date palms, glossy-leaved Golden Leatherferns,
the grey-green rosettes of a pink Aloe,
trundle the hazed cold light of English dawns.

The brothers, hands planted on each cart side,
watch as pickerel blades, date palms' spiked fronds,
the waxy feathered ferns, are caught by breeze;
the running of wooden wheels on the roads
bobbing, animating, the whole display.
In the half-light, through their steady pacing,

to their eyes these swamp and desert plants change
under the breeze. Killarney Ferns stand in
where Leatherferns were. Aloe's rosettes now
appear as dense-flowered orchids; their pink blooms
clustered bell-shapes of Hellaborines. Now
Bog adder's mouths, Hart's-tongue, Wall Rue, stand stark

against blank city skies, ghost the brothers'
eyes and minds as they set to work, taking
chisels between their fingers to pestle
birds, leaf fronds, large German irises, palms,
pickerels, from corbels of Horton rock.
Delicately and exactly, their hands

bring 'a touch to the stone', 3-D clean-edged
shapes that reach into the cathedral-court
spiked with flat, dull, wrought iron decoration.
At dusk, the brothers retire to the mess
with the other hands, slump asleep, dream of
betony, which fends ghosts from Callan graves.

Steven Matthews

"Boughs being pruned"

'Boughs being pruned, birds preened, show more fair;
   To grace them spires are shaped with corner squinches;
Enriched posts are chamfer'd; everywhere
   He heightens worth who guardedly diminishes;
Diamonds are better cut; who pare, repair;
   Is statuary rated by its inches?
Thus we shall profit, while gold coinage still
Is worth and current with a lessen'd mill.'

GERARD MANLEY HOPKINS

LINNÆUS

A Vision of Linnaeus

I saw a youth walking upon the hills
In the breme Lapland morning, while the sun
Now swerving upward (as a swallow turns
That has not rested on the earth) emblazed
The close fur wrapping him with gold that rippled
I' the flying wind: what time I certified
His cap of fox-skin, and his coat of deer:
And, as he walked, how he would stay his step,
Against the unconquered wind to scrutinize
The ground with flowers and rare growths mottled o'er
In that high region; and the rocks and pools
Sucked there by spongy herbage—not as a girl
Culling wild flowers, who looks for these alone,
But taking with a wide glance all that was
As each a limb of one great animal.
For whether it were moss or flower or fern,
Or fungus growth of rottenness, the bare
Bleached jaw-bone of some stag, or wind-bleached rock,
Or raven's wing in rocky cleft, or foot
Of hare the eagle-owl left, nesting close:
Each sang keen notes of one great anthem still,
Of which the dominant (man, in health, disease,
Or death) rang joyous, with a cry that rent
The harmony up through sunny air to heaven.
Grandly he walked, or grander stood, the wind
Passing, and great thoughts passing on more swift
Within him, what the world had been and was;
While in his hand the flower, held listlessly,
I saw he saw not, for his soul was rapt—

As one who has fasted feels a lightness go
Throughout his frame, conversing more with air
Than solid earth, and running seems to fly.
I saw him hovering about that hill
Like an alighted eagle, staring round
A strange world with a glory in his gaze:
A visitant who momently we fear
Even while we gaze may find his task complete,
And merge into the skies in mystery.

JOHN LUCAS TUPPER

The Garden

Except the voice of birds, there is no sound.
  The fruit-trees rustle not: the sultry whirr
Of insects, it is true, is all around,
  Swimming about me with a swoony stir.
The sparrows peck the peaches. Branches green,
  In most admired confusion interlaced,
Look down upon the walks—a shady screen
Of languid coolness; emerald and bright,
  With golden patches of the Sun's warm eye.
Flecked only by the faintest cirrhi white,
  A blank pale blueness reigns throughout the sky.
It might be thus that Isaac Newton paced,
  And saw the apple drop upon the mould.
That apple more than Adam's sure was graced—
  Telling the laws of Nature's Goddess old,
  A revelation vast and manifold.

WALTER DEVERELL

Argonauta

The chickpea slips its coat
          the sheer but significant layer
              falling away,

the nearest you'll get, to her larger egg-case:
flexible-ephemeral. Resilient-fragile. Opal
non-shell, spiral spirit-cloak of this octopus,
which you may mistake as cousin of
          ammonite
             nautilus
         nautilid
   but no.

The paper nautilus is not this.

          Her brood chamber,
      rather than sink her,
         allows buoyancy.

    It is her bright shadow:
tangible, impossible,
      essential.

KELLEY SWAIN

29

Ammonite

I am not body as I began, my matter – its substrates and fleshedness – corroded within the calcifying shell around my shell, the content of my self and soul becoming new and refilling my space with my ghost-image in stone writing, my ridges and runnels important to the hand, my uniqueness shaped and brailled, my place identified my home discovered

STEVEN MATTHEWS

Rorqual

Meaning, "whale with folds",
as if it was a book, burnt rather than read,
lighting the world with unenlightenment.

Now, we try to read them: whales folded,
whales smooth, quarto and duodecimo,
and does it atone for all the shots fired,
is our new translation vigil enough
for each pungent, greasy candle?

KELLEY SWAIN

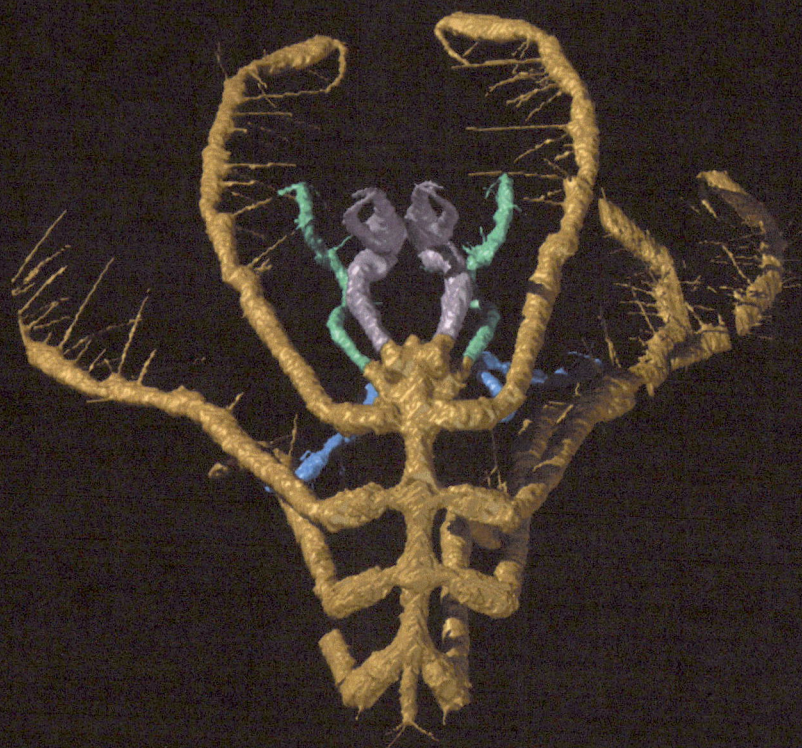

The Sea Spider Speaks
(Rescued From A Lagerstätte)

You took your time but at least you're here;
shut in a ball of stone for millions of years

you've no idea;
I don't mind being sliced by the micron,

exposed in three-D,
made to do back-spins on a computer,

I know I'm just pixels, but everything changes
and it's a better fate than yours who'll be forgotten after death;

when in the Med say 'hi' to the pycnogonids,
I'd strum you a tune but my banjo's lost its strings.

JOHN BARNIE

To the Palaeontologists

Show me the smallest things,
the deep-inside labyrinth
from whence we learned to dive,
the tiny barometer that measures
which way is up, and how far
we can go.
                    Once, we were fishes,
and that left-behind is why we fear
drowning; birds were never ours,
which is why we dream of flight.

KELLEY SWAIN

The Crystallographer
Dorothy Hodgkin

Tessera by tessera,
centuries of Egyptian
sand are eased away.

Patterns of reds, blues,
greens break through
the grains as the brush

strokes across them,
and the mosaic is
gradually unearthed.

A teenager, she crouches
for days shielded
from the sun's blare,

her one-tenth
to scale drawings
emerging into life

as she adds colour
to reveal the tesserae's
repeating pattern.

* * *

Intuitively,
her feet find the rungs
of the wooden ladder,

raised and let down, hand-cranked
on taut wires through pullies.
Thick double-bayed window-

arches and bricked vaulting
hold out nighttime,
hold in silence.

She lowers herself
and her tray
of glass fibres

and shellacked crystals.
She mounts
each minute strand

in the goniometer,
switches up
the unearthed

AC current
drilled through
the lab's two metre

thick wall, then
draws and photos
the x-ray

refractions through
her dyed crystals.
The lattice-work

of atoms flowers
on the photographic plates.
The repeating structure

clarifies for her, except
where birefringence
conjures momently

ghost-spots,
a confusion between
the waves' peaks and troughs,

like the mirages
which shimmered
in sands near Khartoum.

STEVEN MATTHEWS

Corsi Marbles

I *Marbles from Carrara*

Seeming stones of the moon, fine-grained calcites, with bruisings of graphite which might be minor seas rising to the surface. Silvanus, the god of the marble-quarries, and of the burgeoning of woods and the fields, that old man touting his syrinx-pipes, might bask in their reflected light at Colonnata. Lay him offerings of wine and wheatsheaves. Send the wedge-split blocks of Luna down the hill-side slipways, use rollers made from the god's trees to move them to port, lift them with cranes, ship them to Rome, pole them on the shallow barges to the marble-yards. This is a year of triumphs, and Rome gluts on marble. In gratitude, carve a statue of Silvanus, make the marble sing, abrade it with emery from Naxos. Add the god of natural fertility to the panel of the Earth on the Altar of Peace in the Field of War. Have him play to quiet the too-lively twins of Luna marble cradled there. Those twins might be our founders, Romulus and Remus, from this distance: make them calm and ensure our citizens' enduring plenty. Or sound the Luna-Carrara drums of Trajan's Column, see him in rising cartoons sail the Danube, subdue the Dacians. But then watch as the matchstick soldiers, spinning high over his ashes in the column's base, build, for Rome, clad it, and make it shine.

II *Marbles from Claudius' Mountain, Or Teos, Or Simitthus*

Spores in a petri dish, mica flowering from greyish stone. And the 200 camels and 400 donkeys hauled it as 60 foot columns of 240 tonnes each, from staging post to staging post hundreds of miles across the Egyptian desert to the Nile. Workers in Rome dressed the columns, set them up in the emperor's triumphal Forum. Feldspar, and quartz crystals, sometimes captures the Roman light. Or see the greyish blacks mottled with roses and pinks, veined with whites, from Turkey, crossing the Aegean. A bluish-white version decks the Pyramid of Cestius and is inscribed to his glory. Or it is Numidian marble, like a yellowed skin run through with iron veins, set over Caesar the conqueror of North Africa's tomb, cladding the temples of emperors in the capital, and in such as Camulodunum, that dismal outpost. Identifying the marbles in Rome or elsewhere is to read the history of empire, to read the vainglory of our rulers, the excesses of our rich men. Dress your houses in it, set up colonnades of it. Build your palace with each column a commemoration of each of your suppressions. Show your follies. But muse upon marble's varieties, each cut block a different concentration of the earth's upheavals, those tectonic shifts as the African Plate inches North.

STEVEN MATTHEWS

oh, how awful for him,
never willing

to say things clearly.
For if he did,
he would fracture
with faults.

Double Refraction

Everything he said
meant something else

not more, not less,
but both.

He was not safe
unless two-faced.

It was dizzying
to witness:

made her motion-sick,
not sure where she stood.

But oh, how awful for him,
never willing

to say things clearly.
For if he did,

he would fracture
with faults.

KELLEY SWAIN

A Year and a Day

Slow days have passed that make a year,
　Slow hours that make a day,
Since I could take my first dear love
　And kiss him the old way;
Yet the green leaves touch me on the cheek,
　Dear Christ, this month of May.

I lie among the tall green grass
　That bends above my head
And covers up my wasted face
　And folds me in its bed
Tenderly and lovingly
　Like grass above the dead.

Dim phantoms of an unknown ill
　Float through my tired brain;
The unformed visions of my life
　Pass by in ghostly train;
Some pause to touch me on the cheek,
　Some scatter tears like rain.

A shadow falls along the grass
　And lingers at my feet;
A new face lies between my hands —
　Dear Christ, if I could weep
Tears to shut out the summer leaves
　When this new face I greet.

Still it is but the memory
   Of something I have seen
In the dreamy summer weather
   When the green leaves came between:
The shadow of my dear love's face —
   So far and strange it seems.

The river ever running down
   Between its grassy bed,
The voices of a thousand birds
   That clang above my head,
Shall bring to me a sadder dream
   When this sad dream is dead.

A silence falls upon my heart
   And hushes all its pain.
I stretch my hands in the long grass
   And fall to sleep again,
There to lie empty of all love
   Like beaten corn of grain.

ELIZABETH SIDDAL

Summer Dawn

Pray but one prayer for me 'twixt thy closed lips,
    Think but one thought of me up in the stars.
The summer night waneth, the morning light slips
    Faint and gray 'twixt the leaves of the aspen, betwixt the
        cloud-bars,
That are patiently waiting there for the dawn:
    Patient and colourless, though Heaven's gold
Waits to float through them along with the sun.
Far out in the meadows, above the young corn,
    The heavy elms wait, and restless and cold
The uneasy wind rises; the roses are dun;
Through the long twilight they pray for the dawn
Round the lone house in the midst of the corn.
    Speak but one word to me over the corn,
    Over the tender, bow'd locks of the corn.

WILLIAM MORRIS

The Woodspurge

The wind flapped loose, the wind was still,
Shaken out dead from tree and hill:
I had walked on at the wind's will,—
I sat now, for the wind was still.

Between my knees my forehead was,—
My lips, drawn in, said not Alas!
My hair was over in the grass,
My naked ears heard the day pass.

My eyes, wide open, had the run
Of some ten weeds to fix upon;
Among those few, out of the sun,
The woodspurge flowered, three cups in one.

From perfect grief there need not be
Wisdom or even memory:
One thing then learnt remains to me,—
The woodspurge has a cup of three.

DANTE GABRIEL ROSSETTI

A Quiet Place

My friend, are not the grasses here as tall
As you would wish to see? The runnell's fall
Over the rise of pebbles, and its blink
Of shining points which, upon this side, sink
In dark, yet still are there; this ragged crane
Spreading his wings at seeing us with vain
Terror, forsooth; the trees, a pulpy stock
Of toadstools huddled round them; and the flock—
Black wings after black wings—of ancient rook
By rook; has not the whole scene got a look
As though we were the first whose breath should fan
In two this spider's web, to give a span
Of life more to three flies? See, there's a stone
Seems made for us to sit on. Have men gone
By here, and passed? or rested on that bank
Or on this stone, yet seen no cause to thank
For the grass growing here so green and rank?

WILLIAM MICHAEL ROSSETTI

Daphne, The Orchid Mantis

She does it exceptionally well.
She'll do it again: in time,
she'll fledge, and bend;
she will not move for hours on end.

Can she embody patience
if she cannot feel impatience?
Can she embody cruelty

(she eats the cricket live,
starting with its head)

if she cannot feel kindness?

If the trees had no tops
she would keep climbing.

Moulting is an art,
like everything else.

KELLEY SWAIN

Flashman

I.

You need money
and a fear of oranges
where we're going, my boy:
the Silk Trade Route.

Where the river swells like a woman's hips,
and you can't see straight for song:
Oh, let us go! Let us go into our freedom,
we don't want to be faithful,
we just don't want to get caught.

II.

How to build an Empire:
1. Take what catches your eye.
2. Disregard value.
3. Look far from yourself.
4. Think all that glitters is gold.

III.

The only goddamn way
to get some sun round here
is to sink your nails
into exotic flesh:

The satsuma saves me
in the darkest nights of winter.
It's the only star I'll follow,
and I am not a wise man.

KELLEY SWAIN

Who Shot Cock Robin?

Columba P. Palumbus, gumshoe,
presents me with her card, wants
to ask questions, won't say why;
that Scarlet Ibis, she indicates,
that Kakapo; that Maroon-Breasted
Crowned Pigeon, bit of a mouthful,
known as 'Dizzie'; that Snowy Owl
rising in a flurry of icy feathers;
she looks at me inquiringly; Corvus
Corone taken 6 June '25, with a
nest of demanding mouths;

(when Mounteney Jephson killed
a Crested Crane, 1888, 'its mate
flew about uttering the most mourn-
ful cries', 'one felt,' he said, 'like a
murderer', omitting 'almost');
Palumbus writes this down, snaps
shut her notebook with a frown.

JOHN BARNIE

Let's Do It

Recreate a dodo from its DNA
so it could waddle in an aviary

or among the legs of tourists on Mauritius;
call it the resurrection bird;

isn't it ugly, what do you feed it on,
people might say

strolling to the beach café
where a blue sky meets a blue sea

dreaming of paradise
until they die.

JOHN BARNIE

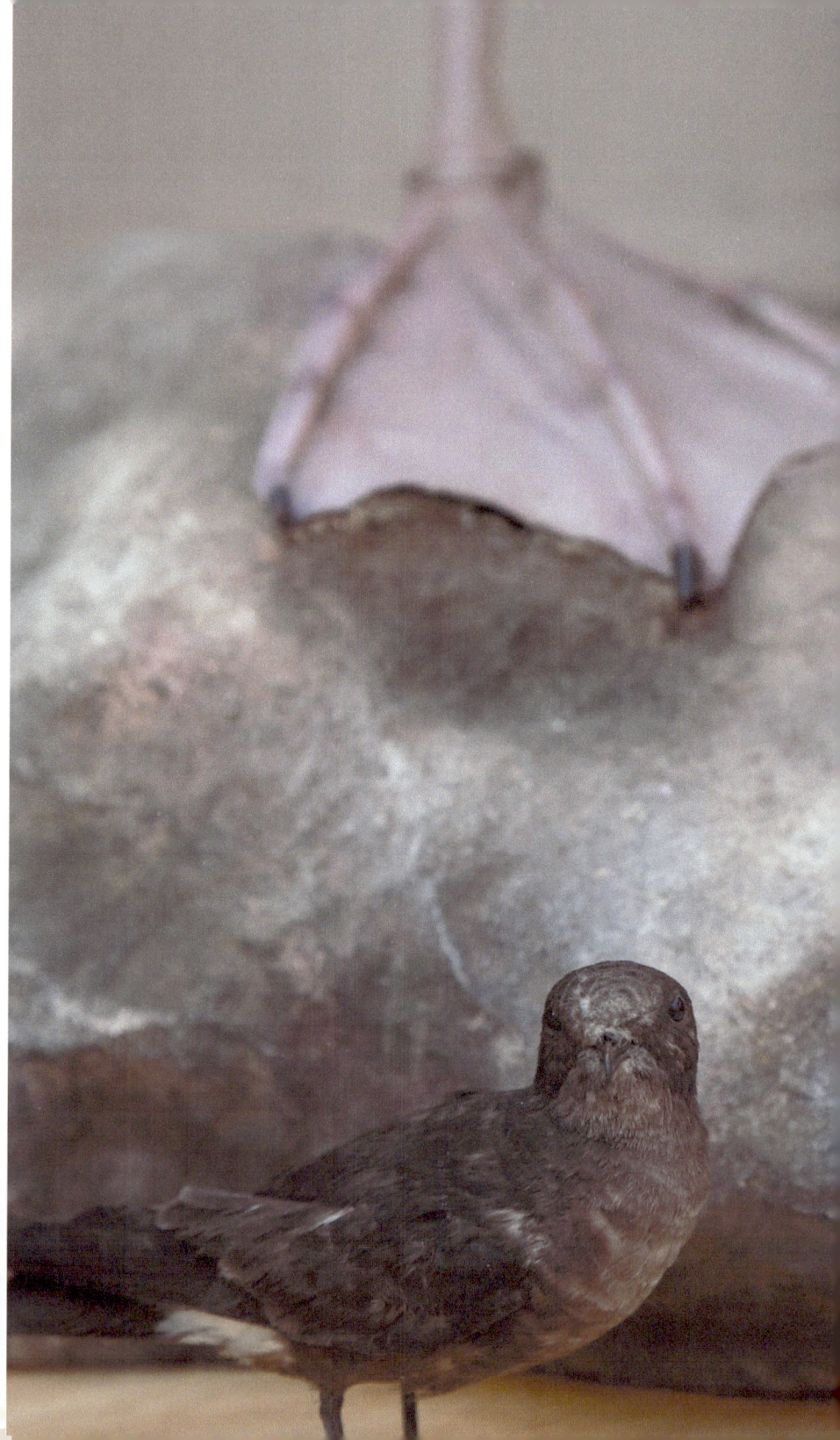

The Wandering Albatross
and the Storm Petrel

Just because they are remotest cousins,
They have posed them here in the same glass case;
Stood like a couple weighing their options,
They gaze to horizons far from this place.

Ghosting the Capes with massive mud-white wings
This handsome storm-scorner wandered at will.
Now stuck on its rock, quizzical at things,
It seems lost, heavy, tired of holding still.

Not so its much littler companion,
Undistinguished in all ways as it is;
It stands perky and sure at its station,
Its short beak eager for squid or for fish.

Perhaps the Poet's bird, the albatross,
Has spent too long roaming the wind's quarters:
Its remote cousin, homing at storms' toss,
Should teach it how to walk upon waters.

STEVEN MATTHEWS

Shadows

Descend on a wobbling ladder of bones into the past?
all we can do is peer over hummocky grass and pink thrift
at what's going on now; give this plesiosaur a pat on the head,
it won't be the head that snaps and bites you with its
jawful of needles, paddling round with one remaining paddle

on the Museum researcher's floor, all its bones disarticulate,
the head not ready yet, but spitting teeth in the clay; something
the Time Traveller might have seen at the end of the world
rolling off a sand bar under a giant red sun; T. rex, too,
the bachelor uncle scaring little children with a roar; they

know it's not real but are scared just the same; look how it's
    running
but never leaves its stand, with tiny tucked-in forelimbs;
it could be grandma's nemesis the wolf in Red Riding Hood,
but it's never never going to hurt you, nor the plesiosaur
thrashing in circles with its paddle, in a sea that's gone forever.

JOHN BARNIE

Emblems

I lay through one long afternoon,
   Vacantly plucking the grass.
I lay on my back, with steadfast gaze
   Watching the cloud-shapes pass;
Until the evening's chilly damps
   Rose from the hollows below,
   Where the cold marsh-reeds grow.

I saw the sun sink down behind
   The high point of a mountain;
Its last light lingered on the weeds
   That choked a shattered fountain,
Where lay a rotting bird, whose plumes
   Had beat the air in soaring.
   On these things I was poring:—

The sun seemed like my sense of life,
   Now weak, that was so strong;
The fountain—that continual pulse
   Which throbbed with human song:
The bird lay dead as that wild hope
   Which nerved my thoughts when young.
   These symbols had a tongue,

And told the dreary lengths of years
   I must drag my weight with me;
Or be like a mastless ship stuck fast
   On a deep, stagnant sea.
A man on a dangerous height alone,
   If suddenly struck blind,
   Will never his home path find.

When divers plunge for ocean's pearls,
  And chance to strike a rock,
Who plunged with greatest force below
  Receives the heaviest shock.
With nostrils wide and breath drawn in,
  I rushed resolved on the race;
  Then, stumbling, fell in the chase.

Yet with time's cycles forests swell
  Where stretched a desert plain:
Time's cycles make the mountains rise
  Where heaved the restless main:
On swamps where moped the lonely stork,
  In the silent lapse of time
  Stands a city in its prime.

I thought: then saw the broadening shade
  Grow slowly over the mound,
That reached with one long level slope
  Down to a rich vineyard ground:
The air about lay still and hushed,
  As if in serious thought;
  But I scarcely heeded aught,

Till I heard, hard by, a thrush break forth,
  Shouting with his whole voice,
So that he made the distant air
  And the things around rejoice.
My soul gushed, for the sound awoke
  Memories of early joy:
  I sobbed like a chidden boy.

THOMAS WOOLNER

"Yet With Time's Cycles Forests Swell"

THOMAS WOOLNER, 'EMBLEMS'

If we kept in our minds
fishes' fins could be wings
raising all up into
a chemical soup sky;
or that a dull granite
boulder could shift with ice
the hundred miles from Shap
to Filey just like that –

or if we imagined
waking into bee-less
silences in spring-times,
worried what we would eat –

if only we'd conceived
of a fossil sepia
sketch of a dinosaur
in ink as old as it;
or if everywhere
raised cathedrals of light
and scope uplifting life,
showing us what life is –

if only we'd foreseen
that pesticides might flow
from fields and suffocate
Hughes's salmon and pike;
or known that reserpine
from shrubs reduces stress
and its rainforest homes
are already half gone –

if we'd seen razor-bills
surfboarding on branches
to behead frogs downstream;
or a Cleopatra
flexing yellow wing-tips
and raising itself to
blue Banbarian skies
like a coloured flickbook –

or we'd understood that
whatever in a day
comes into being dies
from the world that same day –

maybe we'd change our thoughts.
Maybe we would think.

STEVEN MATTHEWS

Here's A Modest Apology
for Anthropo-Pithcology

Why not the Monkey rehabilitate?
Their courtesy why not reciprocate?
Watching them ape the Man we've stood agape; —
'Tis surely time to try and man the Ape.

JOHN ADDINGTON SYMONDS, SR.

An Old Gordian Knot

Part of the whole that never can be known,
  Is this poor atom that we call our world;
  Part of this part amid confusion hurled
  Is man, an idiot on a crumbling throne:
Yea, and each separate soul that works alone,
  Striving to pierce the clouds around him curled,
  Gasps but one moment in the tempest whirled,
  And what he builds strong Death hath overthrown.
How shall this fragment of a waif, this scape
  In the oblivion of unreckoned years,
  This momentary guest of time, this ape
That grins and chatters amid smiles and tears,—
  How shall he seize the skirts of God and shape
  To solid form the truth that disappears?

JOHN ADDINGTON SYMONDS, JR.

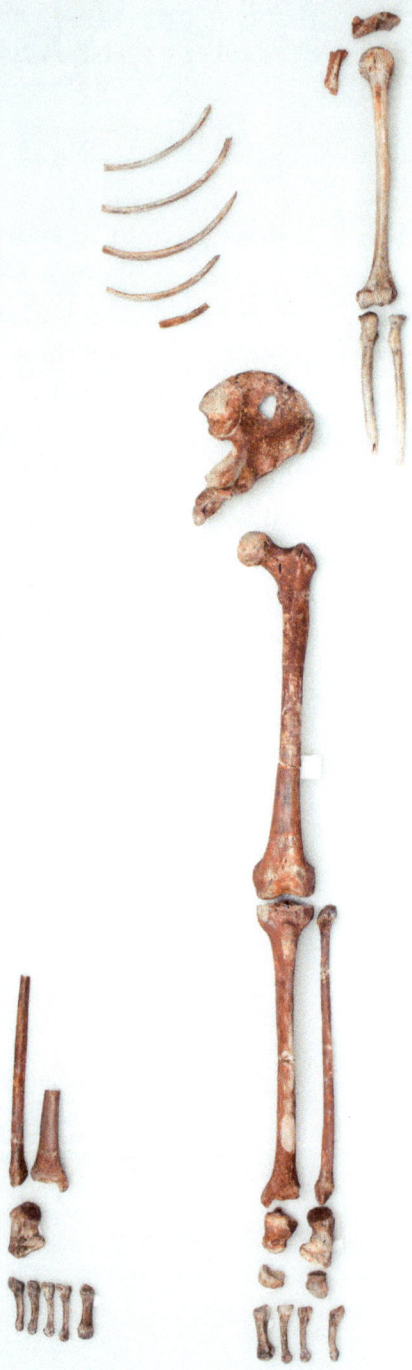

A Welshman Looks at the Red Lady

You could line up skulls in these cases, make a xylophone,
play a clacketty-clocketty kind of music, get the Red Lady up
one-leg dancing, bit of a shambles but in the groove; or leave

her-him to be educational, not caring much I should think;
H. habilis, H. erectus, H. ergaster, count them out until
you come to us and the Red Lady-man, Linnaeus off-piste

when he named us sapiens (nothing like being hopeful)
though sapientiae capax floats out of the dictionary
as a better explanation of how we arrived, So-and-So the hominid,

driving around in cars, drinking in bars, as waves whump
and the Promenade is closed by police in yellow jackets;
we might have stayed habilis, we might have stayed erect,

technology not the wedge, not the hammer-headed blow
that split us from the past until it's a question of 'Dance
Little Lady, Dance-Dance-Dance' to the ticking of the clocks.

JOHN BARNIE

Fibonacci Crystals

So
It is
Incomplete
Unfortunately
The hesitant progress of life,
The limitations constantly being adapted
Into new possibilities
Bright new butterflies
Launch into
Clear wide
Skies

So
Mobile
Shelving rests
The upper jaw bone
Of the amphibious rhino
That died in the South African bushland at the time
The massed nations at Waterloo
Died in their thousands
Their jaws pressed
Down in
Mud

So
Poems
Are crystals:
They take the life lived,
Narrow it, diffract it, project
Its pattern on the mind's photographic plates, atoms
Integrated, othered, as forms
Thrown beyond the shapes
Held within
Its voice
Breaths

STEVEN MATTHEWS

PYTHONIDAE

The Spirit Collection

There's a foetus suspended in a dance,
I'd call it 'Merrylegs',

a spinal column, human too,
with the rope of nerves severed from the brain,

a child's fur coat from which a monkey's eyes peep out
teeth exposed by a curling of the lip;

switch off the lights, tiptoe away,
leave the foetus to its dance, the monkey to its peeping,

the spinal column making the brain's last call
asking how it came to end like this.

JOHN BARNIE

The Spirit Room

We, the lung, heart-apex,
gut, left eyeball, left nostril,

caecum, lung and diaphragm
of kiwi, emu, ostrich,

suspended here at the other end
of the world from where we were,

have lost our voice
have lost our stories

to the silences of formalin
or of Serventy's Fluid,

greyish, bleaching,
neither male nor female,

afloat in
separate hemispheres.

* * *

When the Maenads
tore Orpheus

and cast his bloodied limbs
into the Helicon or Turkish Hebrus,

his mouth sang mournfully
across waters migrating

from river to sea to Greek Lesbos.
If you attend,

we unlikely migrants,
mouthless yet eloquent,

can sing you unhampered truths
of lostness and purpose

through the walls of
our glass-caged world.

STEVEN MATTHEWS

Lyme

What if the real gods
are the deer ticks,
not the deer?

That hot destruction,
a legacy of cramped pain
bloating the joints, fingering

crevices in the brain like a monk
at his prayer beads. Better
to know you've been bitten:

the red bull's-eye reminder
of quickened mortality, predator
turned prey, gives a shot
at salvation. What if

the real gods are the little ones,
not impossible to see,
but too small?

KELLEY SWAIN

Cockroaches On My Mind

In the House of Dust they're ready to scuttle
with a chitin rustle and gleam through bottlenecks
to radiate across horizons of millions of years,
not like humans, doubling back on ourselves in cities,

staring from a high-rise at billions of our kind,
an infestation nearly at an end; 'bottleneck, bottleneck,
who will survive', goes the rhyme a mother sings
as she rocks the cradle in a considering way;

leave that to the cockroaches who prickle the skin
with a delicate stipple of multiple feet, antennae
feelingly trying to tell where they are and whether
this might be something to eat; I like them, didn't

think I would; let them carry on; 'your time now,
our time after a while', the words of an old song
called 'Bottleneck Blues'; let's put the record on,
invincible, as we say we are, in the House of Dust.

JOHN BARNIE

The Lay of the Trilobite

A mountain's giddy height I sought,
    Because I could not find
Sufficient vague and mighty thought
    To fill my mighty mind.
And, as I wandered ill at ease,
    There chanced upon my sight,
A native of Silurian seas,—
    An ancient Trilobite.

So calm, so peacefully he lay,
    I watched him e'en with tears.
I thought of Monads far away,
    In the forgotten years.
How wonderful it seemed, and right,
    The providential plan,
That he should be a Trilobite,
    And I should be a Man!

And then, quite natural and free,
    Out of his rocky bed,
That Trilobite he spoke to me,
    And this is what he said:
'I don't know how the thing was done,
    Although I cannot doubt it;
But Huxley—he if anyone
    Can tell you all about it:—

How all your faiths are ghosts and dreams,
    How, in the silent sea,
Your ancestors were Monotremes—
    Whatever these may be,—
How you evolved your shining lights
    Of wisdom and perfection,
From Jelly-fish and Trilobites,
    By Natural Selection.

You've Kant to make your brains go round,
    And Carpenter to clear them,
And Mathematics to confound,
    And *Mr. Punch* to cheer them.
The native of an alien land
    You call a man and brother,
And greet with pistol in one hand,
    And hymn-book in the other!

You've Politics to make you fight,
    And utter exclamations,
You've cannon, and you've dynamite
    To civilise the nations,
The side that makes the loudest din
    Is surest to be right,
And oh, a pretty fix you're in!'
    Remarked the Trilobite.

'But gentle, stupid, free from woe,
    I lived among my nation,
I didn't care, I didn't know,
    That I was a crustacean.
I didn't grumble, didn't steal,
    I *never* took to rhyme,
Salt water was my frugal meal,
    With carbonate of lime.'

Reluctantly I turned away,
    No other word he said;
An ancient Trilobite he lay
    Within his rocky bed.
I did not answer him, for that
    Would have annoyed my pride,
I merely bowed, and touched my hat,
    But in my heart I cried:—

'I wish our brains were not so good,
    I wish our skulls were thicker,
I wish that Evolution could
    Have stopped a little quicker.
For oh, it was a happy plight
    Of liberty and ease,
To be a simple Trilobite
    In the Silurian seas!'

MAY KENDALL

OPHICALCITE (SERPE...
CONNEMARA, IREL...

Oxford

Go to Oxford for a real souvenir:
when you need to remember magic.
                    Go, on days
when drowsy voluptuous roses
nod approval at quotes
                    you misremember,
go, watch Worcester's lakeside heron
practise Silurian meditation.

If hope is a thing,
it is a thousand languages,
                    a thousand
libraries, chapels, collections –
architectural, then emotional,   parallelism
which connects, whilst going on   indefinitely.

That bug is not a bug   it is a type specimen.
That gem is not a gem   it is a curse.
This poem is not a poem   it is a superstition.

KELLEY SWAIN

Notes

INTRODUCTION

The quotations from Acland are taken from the *Oxford Archi-
tectural Society: Reports of Meetings from July 1853, to May 31,
1856* and from Henry W. Acland and John Ruskin, *The Ox-
ford Museum*, new ed., published in 1893. The essays by the
Pre-Raphaelites which are quoted are Frederic George Stephens
[as John Seward], 'The Purpose and Tendency of Early Italian
Art', published in *The Germ* in 1850; William Michael Ros-
setti, 'Pre-Raphaelitism', published in the *Spectator* in 1851;
and Stephens, 'The Oxford University Museum', published in
Macmillan's Magazine in 1862. For a full account of the build-
ing of the Museum, see Frederick O'Dwyer, *The Architecture of
Deane and Woodward* (Cork: Cork University Press, 1997), pp.
152-283; for a discussion of the place of science in Pre-Rapha-
elite art theory, see John Holmes, 'Pre-Raphaelitism, Science,
and the Arts in *The Germ*', *Victorian Literature and Culture*, 43
(2015), 689-703.

THE O'SHEAS

James and John O'Shea, together with their nephew Edward
Whelan, carved the botanical and zoological stonework at the
Oxford Museum in the 1850s and early 1860s.

"BOUGHS BEING PRUNED"

This fragment was written in the mid-1860s while Hopkins
was a student at Oxford.

A Vision of Linnaeus

First published in *Poems by the Late John Lucas Tupper*, ed. by W. M. Rossetti (1897). The Swedish naturalist Linnaeus established the first comprehensive taxonomy of the natural world in the eighteenth century. Through Dante Gabriel Rossetti, Tupper got the commission to carve the statue of Linnaeus at the Museum. This poem forms a double work of art with the statue.

The Garden

This poem is preserved in an unpublished manuscript edited by William Rossetti in the Huntington Library in California.

The Crystallographer

Dorothy Hodgkin conducted her pioneering work in x-ray crystallography at the Oxford Museum. She was awarded the Nobel Prize for Chemistry in 1964 and deciphered the structure of insulin in 1969.

A Year and a Day

This poem dates from around 1857. It exists in different forms in various manuscripts. The text given here is reprinted from the online text prepared by Ian Lancashire at the Department of English at the University of Toronto (https://tspace.library.utoronto.ca/html/1807/4350/poem2729.html).

Summer Dawn

Reprinted from *The Defence of Guenevere, and Other Poems* (1858).

The Woodspurge

This poem dates to around 1856, when Rossetti was acting as a consultant on the design and decoration of the Museum. It is reprinted from *Poems* (1870), where it was first published.

A Quiet Place

Reprinted from *The Germ* (1850).

Emblems

Reprinted from *The Germ* (1850).

Here's a Modest Apology For Anthropo-Pithcology

This poem was included by Charles Daubeny, one of the founders of the Museum, in his book *Fugitive Poems Connected with Natural History and Physical Science*, published in 1869.

An Old Gordian Knot

This sequence of eight sonnets was first published in *New and Old: A Volume of Verse* (1880).

A Welshman Looks at the Red Lady

The Red Lady of Paviland was discovered in South Wales in 1823 by Oxford University's first Reader in Geology, William Buckland. Buckland interpreted the skeleton, dyed with red ochre, as the remains of a woman from the period of the Roman occupation of Britain. More recent analysis has revealed that the bones are those of a man who lived approximately 33,000 years ago.

First published in *Punch* in 1885. The text here is reprinted from the first printing. Thomas Henry Huxley was the leading public advocate for Darwinism and formal science more generally from the early 1860s onwards. When this poem was published he was President of the Royal Society and Dean of the Normal School of Science in South Kensington in London. William Carpenter was the leading British expert on physiology and neurology, and a pioneer in the field of psychology.

Biographies

JOHN BARNIE is from Abergavenny in the Usk Valley on the edge the Black Mountains. He taught for a number of years at Copenhagen University before giving up academic life and returning to Wales to work as editor of the cultural and political magazine *Planet*. He has published 15 collections of poetry as well as fiction and essays. His latest book is a memoir, *Footfalls in the Silence*; a new collection of poems, *Wind Playing with a Man's Hat*, will appear at the end of 2016 (both from Cinnamon Press). He performs with the blues and poetry band Hollow Log.

SCOTT BILLINGS works on the Oxford University Museum of Natural History's digital engagement activity in a role that includes photography, videography, and design, as well as exhibition development and media communications. He also teaches a digital camera course at the Oxford School of Photography and has a background in design journalism.

WALTER DEVERELL was born in 1827. He studied painting at the Royal Academy, where he became a friend of Dante Gabriel Rossetti. Through Rossetti he became one of the close Pre-Raphaelite circle, though he was never a member of the Pre-Raphaelite Brotherhood itself. His best-known paintings depict scenes from Shakespeare's comedies. Deverell died of Bright's disease in 1854 when he was only twenty-seven.

JOHN HOLMES is Professor of Victorian Literature and Culture at the University of Birmingham. His books include *Darwin's Bards: British and American Poetry in the Age of Evolution* and the edited collection *Science in Modern Poetry: New Directions*. He is currently writing a book on the Pre-Raphaelites and science.

GERARD MANLEY HOPKINS was born in 1844. He had a life-long interest in the natural sciences, including astronomy and natural history. He studied classics at Balliol College, Oxford, in the 1860s. His fragmentary poem "Boughs being pruned" dates from this period. In 1866 he converted to Roman Catholicism, later training to be a Jesuit priest. He died in 1889. His poetry remained in manuscript during his lifetime, only becoming famous in the twentieth century when it was published by his friend and Oxford contemporary the Poet Laureate Robert Bridges.

MAY KENDALL was born in 1861. She was a student at Somerville College, Oxford. In the 1880s and 1890s she published several comic poems engaging with science. Kendall became an active campaigner for women's rights and social justice. She died in 1943.

STEVEN MATTHEWS is a poet and critic who was raised in Colchester, Essex, and now lives in Oxford. His poetry collection *Skying* appeared from Waterloo Press, UK, in 2012. He has been a regular reviewer for journals including the *TLS*, *Poetry Review* and *London Magazine*. He has been Poetry Editor for *Dublin Quarterly Magazine*. As a critic, Steven Matthews has published books on a wide range of twentieth- and twenty-first century poetry in English, including writing on Yeats, T.S. Eliot, Les Murray and contemporary Irish poetry.

WILLIAM MORRIS was born in 1834. After studying classics at Exeter College, Oxford, he began an apprenticeship in 1856 with the Gothic architect G. E. Street, who had advised on the building of the Oxford Museum. In 1857 Morris worked with his friends Dante Gabriel Rossetti, Edward Burne-Jones and others on a project to decorate the Oxford Union Society debating chamber, built by Benjamin Woodward, the architect of the Museum. His first book of poems was published the following year. Morris went on to be one of the pioneers of English Socialism and the most influential decorative artist of his age, inspiring the Arts and Crafts Movement. He died in 1896.

DANTE GABRIEL ROSSETTI was born in 1828. He was one of the leaders of the Pre-Raphaelite Brotherhood, a group of seven young artists including John Everett Millais and William Holman Hunt, which was launched in 1848. Their aim was to revolutionize Victorian art, rejecting the conventions taught at the Royal Academy, and looking instead to the practices of medieval art and modern science. Both a poet and a painter, Rossetti was the charismatic visionary of the group, who went on to inspire a second circle of artists and writers including Morris and Burne-Jones. He was consulted by Woodward, Ruskin and Woolner on their work for the Oxford Museum. 'The Woodspurge' dates from around this time. When he died in 1882, he was widely recognized as one of the most brilliant and influential artists and poets of his time.

WILLIAM MICHAEL ROSSETTI was the younger brother of Dante Gabriel Rossetti. Born in 1829, he was the only member of the Pre-Raphaelite Brotherhood who was not a trained artist. His role was to document the P.R.B.'s activities, keeping the Pre-Raphaelite Journal, and to edit its manifesto magazine *The Germ*, which ran to only four issues in 1850. A civil servant, he had a parallel career as a prolific and respected art critic. He outlived all the other Pre-Raphaelites, not dying until 1919.

LIZZIE SIDDAL was born in 1829. She met Walter Deverell in 1849 when she was working as a milliner in London. As a result of this chance meeting, she became the most famous Pre-Raphaelite model, modeling for John Everett Millais's *Ophelia* and several paintings and sketches by Dante Gabriel Rossetti. Rossetti and Siddal became lovers and artistic collaborators, their art and poetry developing alongside one another across the decade. Siddal's paintings were admired by John Ruskin, who became her patron. As she was by this time persistently unwell, Ruskin arranged for his friend Henry Acland, who had been instrumental in the campaign to build the Oxford Museum, to be her doctor. Siddal was asked to contribute designs for decorative carvings at the museums in Dublin and Oxford that Benjamin Woodward was building, though in the end they were not used. In 1860, after a turbulent and protracted courtship, Siddal and Rossetti finally married. She died two years later of an overdose of laudanum.

KELLEY SWAIN has lived in London for nearly ten years, working at the crossroads of poetry and science. She is the author of three books of poetry, a novel, and a memoir, and editor of two poetry-and-science anthologies, one of which was the culmination of her residency at the Whipple Museum of the History of Science at the University of Cambridge. She holds an MSc in Medical Humanities from King's College London, contributes regularly to the reviews pages of *The Lancet* journals, and was artist-in-residence at Duke University, helping to launch their new Health Humanities Laboratory, in autumn 2016.

JOHN ADDINGTON SYMONDS (1807-1871) was a leading Victorian doctor. His son, also JOHN ADDINGTON SYMONDS (1840-1893), studied at Balliol College, Oxford, from the late 1850s. The younger Symonds described Charles Darwin's discovery of evolution by natural selection as 'a theory which startled the world'. As well as being a poet, he was a prominent historian, classicist and literary critic. He is remembered today as a pioneering campaigner for homosexual rights.

JOHN LUCAS TUPPER was born in the mid-1820s. In the 1840s he studied sculpture at the Royal Academy, where he became a friend and mentor to the Pre-Raphaelite Brotherhood. Tupper worked as an anatomical illustrator at Guy's Hospital from 1849 to 1863, before going on to teach drawing at the University of London and Rugby School. He died in 1879. His statue of Linnaeus at the Oxford Museum is his major work of sculpture and the most thorough-going application of the Pre-Raphaelite principal of truth to nature in that medium. His poems, including the companion piece he wrote for this statue, were published in 1897, long after his death, by William Michael Rossetti.

THOMAS WOOLNER was the only member of the Pre-Raphaelite Brotherhood who was a sculptor rather than a painter. Born in 1825, he first made an impression with his impish statuette of Puck from *A Midsummer Night's Dream*, which was exhibited in 1847. Woolner worked closely with his Pre-Raphaelite colleagues, carrying their artistic principles over from painting and poetry into sculpture. After beginning his career making portrait busts and medallions, his first major commission for public sculpture was the statue of Francis Bacon at the Oxford Museum, followed a few years later by the memorial statue of Prince Albert. Beginning with his work at the Museum, until his death in 1892, Woolner was one of England's leading sculptors.

.